Captured Weapons and Equipment of the German Wehrmacht

1938-1945

Two Fiat-Spa Pavesi P 4-110 heavy artillery tractors, one towing a 75 mm Field Cannon 237 (i) (Italian 75/27 M.06); the second tractor has an ammunition trailer in tow. They were taken over by a Navy command at a support point in Greece and photographed in September 1943.

Wolfgang Fleischer

Schiffer Military/Aviation History
Atglen, PA

Bibliography

D 41/2, Gebrauchsanleitung für russisches Gerät, Berlin 1941.

D 50/1-15, Kennblätter fremden Geräts, Berlin 1941.

D 51/2, Gebrauchsanleitung für englisches Gerät, Berlin 1941.

D 98, Waffen des Auslandes, Berlin 1937.

H.Dv. 119/955, Vorläufige Schusstafel für den 8,14 cm Granatwerfer 278 (f)-frz. 27/31. Berlin 1942.

H.Dv. 481/134, Merkblatt für die Munition des 8 cm Granatwerfer 36 (t), Berlin 1942.

Merkblatt 29/14, "Russische Spreng- und Zündmittel, Minen und Zünder, Berlin 1943.

Merkblatt Waffen der Roten Armee, Posen 1944.

Author's notes on salvage of a turret from a Panzerkampfwagen 38 (t) in Frankfurt/Oder, Dresden 1985.

Author's notes on salvage of fragments of a 22 cm Mortar 531 (f)- frz. 16 TR, Dresden 1988ff.

Fischer, K, Waffen- und Schiesstechn. Leitfaden für die Ordnungspolizei, Berlin 1943.

Hahn, F., Waffen und Geheimwaffen des deutschen Heeres 1933-1945, Vol. I & II, Koblenz 1986 & 1987.

Heiden, H., Gewehre frei! Weg und Ruhm der Maschinengewehr-Waffe, Berlin 1938.

Hoppe, H., Die 278. ID in Italien 1944/45, Bad Nauheim 1953.

Kisser, H., Der Deutsche Volkssturm 1944/45, Berlin & Frankfurt am Main 1945.

Koch, F., Beutepanzer im Ersten Weltkrieg, Wölfersheim-Berstadt 1994.

Leeb, E., Aus der Rüstung des Dritten Reiches, Frankfurt am Main 1958.

Ludendorff, E., Meine Kriegserinnerungen 1914-1918, Berlin 1921.

Mueller-Hillebrand, B., Das Heer 1933-1945, Vol. I & II, Frankfurt am Main 1954.

Novah, J., Befestigung in der Gegend von Králicky, Zamberh 1994.

Oswald, W., Kfz. und Pz. der Reichswehr, Wehrmacht und Bundeswehr, Stuttgart 1990.

Paul, W., Panzergeneral Walther K. Nehring, Waffen der Roten Armee, Posen 1944.

Schlauch, W., Rüstungshilfe der USA 1939-1945, Siegburg-Niederpleis 1962.

American 2.5-ton Studebaker US 6-U2 (6x6) trucks came into German possession chiefly on the eastern front. This picture was taken in 1943.

Translated from the German by Ed Force

Copyright © 1998 by Schiffer Publishing, Ltd.

Printed in China.
ISBN: 0-7643-0526-3

This book was originally published under the title,
Waffen Arsenal-Beutewaffen und -Gerät der Deutschen Wehrmacht 1938-1945
by Podzun-Pallas Verlag.

We are interested in hearing from authors with book ideas on related topics.

COVER PICTURE:
Infantry Tractor UE 630 (f)-frz. Chenilette with 28/32 cm launching frame. The vehicle is part of the collection at the Militärhistorisches Museum der Bundeswehr, Dresden.

Published by Schiffer Publishing Ltd.
4880 Lower Valley Road
Atglen, PA 19310
Phone: (610) 593-1777
FAX: (610) 593-2002
E-mail: Schifferbk@aol.com
Please write for a free catalog.
This book may be purchased from the publisher.
Please include $3.95 postage.
Try your bookstore first.

In 1942, the 4th battery of the II. Unit of Army Artillery Regiment 84, to replace two 24 cm cannons, received three 19/4 cm 485 (f)-frz. GPF cannons on self-propelled mounts. They saw service with the Army Group North.

Introduction

Weapons, ammunition, military vehicles and equipment, in short, items of military technology, have a high material value. Their production causes considerable burdens for any state, and often enough, only that factor limits the expansion of its armed forces. If one side was able to capture weapons and equipment in the process of military activities and put them into use, they could thus strengthen their military presence considerably. The opponent was also weakened in the process.

So it is not surprising that captured war materials were always put into use in all times. There were only differences as to the intensity with which this intention was realized. This is understandable too, for the military developments in various countries were often based on very different tactical and technical concepts. Thus they could not be taken over as they were by other armies. There were many reasons for this: the military doctrine, the training and capability of the officers and soldiers, safety considerations, etc.

The technological basis for the production of war materials has varied from one country to another. This resulted in certain qualitative characteristics which, in terms of reuse, meant that certain characteristics, such as the choice of calibers for artillery weapons, made the reuse of captured materials by the other side difficult or even impossible. As a rule, this was limited to a short period of time until ammunition or important spare parts were available.

In World War II, captured war materials were put back into use by all the forces involved in the war. There is much proof of this. One particular case, though, was that of the German Wehrmacht. Captured weapons, ammunition, vehicles and equipment, and even uniforms, were used in great numbers and over a large period of time. Their utilization was of such great importance for the maintenance of the fighting forces, particularly the army, that the production of ammunition and the most important spare parts had to be carried on, naturally by utilizing the production capacities in the occupied countries.

It is not the author's purpose in this Weapon Arsenal book to portray the entire extent of these interesting and complex problem situations, divided into individual genres. That will be reserved for special works that, for example, already exist on the topics of tanks and other armored vehicles. The purpose here is to depict some general characteristics of the utilization of captured war materials by the German fighting forces between 1939 and 1945.

This 77 mm anti-aircraft gun was produced by using captured French 75 mm field cannons made in 1897. The picture was taken in 1918.

Already in World War I—Born of Need

Oberstleutnant H. G. Garcke dealt in an essay, published in 1936, with the "sins of omission in the military armament of Germany before the war". In it he propounded numerous reasons that had resulted, during the four years of World War I, in a sometimes threatening shortage of all kinds of war materials. Among them, according to his conception, were the completely insufficient cooperation of the military offices in the area of armaments, as well as the neglect of economic preparations for war in connection with false concepts of the course and extent of the war, and finally "the not sufficient evaluation of technology". These cases of neglect could no longer, or only quite insufficiently, be disposed of in a war, with all its pressures on Germany. Indeed, the production of weapons and ammunition increased enormously and attained dimensions hitherto unimagined. Erich Ludendorff, First Quartermaster General and Chief of the General Staff of the Field Army in the war, wrote of this in his war reminiscences in 1918: "Primarily, we needed more artillery, ammunition and machine guns, as well as greater numbers of many other things. The artillery was needed not only for new arming but also for re-arming, in order to replace older types with newer ones. and finally to replace the very many that were put out of service. In the battles before Verdun and on the Somme, we had very heavy losses not only from enemy fire, but also as a result of the demands on the materials from our own increased rate of fire."

The fighting forces of the Entente were always far ahead of the Germans in the realm of military armaments. The troops at the front became very aware of this. Sometimes more, sometimes less, but the tendency increased. Faults appeared above all in the types of weapons whose importance before the war had been estimated wrongly. Here captured weapons and supplies of ammunition were made use of and helped the cause in many cases. Here are some examples:

1. In 1914 the German Army had something more than 2000 machine guns with their crews in the field (in all, 4900 were on hand). Soon these weapons played a central role in defensive fighting; the need for them multiplied greatly, and the industry could not meet the need at once. In order to make up for the losses, the troops were given captured machine guns. In particular, several hundred of the [p. 7] Russian 1910 model machine gun were available after the German victories at Tannenberg and the Masurian Lakes. Among others, the motorized MG unit of the German Army High Command in Belgium was supplied with these captured guns. Machine guns made in France (circa 1914) were used in large numbers, with makeshift mounts, as anti-aircraft guns. In 1916 the Royal Prussian War Ministry issued a publication, "Captured Machine Guns". It was meant to make it easier for the troops to use these desired weapons.

This Russian fortress cannon, dating from 1877, was put to use by the Germans in World War I. The gun, weighing 5856 kilograms, had a caliber of 203.1 mm and could fire up to 10,000 meters.

Captured French 1914 machine guns were set in such makeshift mounts by German units for anti-aircraft use. This picture was taken in 1917.

Below: The Russian M. 1910 machine gun, here in its oldest version with smooth cooling mantle, was used in large numbers by the German Army in World War I.

2. Soon after the war broke out, the active service of flying troops for the Entente demanded effective anti-aircraft weapons for the German Army. They had too few anti-aircraft guns of their own. The obsolete C/73 and C/73/91 field cannons also did not meet the needs, what with their insufficient ballistic performance and the makeshift character of their mounts. The industry could not immediately meet the troops' needs for large numbers of effective anti-aircraft guns because of the large-scale production of field guns. The troops made do by adapting captured French and Russian field guns as anti-aircraft guns. In 1918 there were 394 or 608 of them. This was a quarter of the whole supply of guns for the anti-aircraft troops.

3. During World War I, the artillery became the dominant factor on the battlefields. The need for heavy artillery, the shortage of ammunition, as well as the wealth of captured guns resulted in the establishment of immobile batteries for service in quiet sectors of the front. In the history of the former Saxon 2nd Foot Artillery Regiment No. 19, one can read: "Until June 1916, eight such batteries were sent to the front by the regiment". Some had Japanese howitzers, others Russian 15 cm cannons and 20 cm howitzers. The gunners could often use the guns only under difficult conditions. The batteries did not have a long life expectancy; gradually they were supplied with modern German guns.

4. In the course of the war it became evident that tanks, despite many insufficiencies, were an effective alternative to the use, costly in every sense, of artillery, in order to overcome the machine gun as the dominant defensive weapon. The German military command realized this too late. Thus the development of German tanks was delayed. In order to obtain a sufficient number of tanks in a short time, for use in the offensive operations in the spring of 1918, captured British tanks were overhauled and put into service. By April 1918 there were 150 of them. Four "Sturmpanzerwagenabteilungen" (Beute) were ready by the end of May 1918, and further units were planned. Captured tanks remained in use by the German troops until the end of the war.

Further examples in the realm of arming with tanks, other armored vehicles, infantry guns and other materials could be cited. They all show that the reuse of captured war materials by the Germans was more than a spatially and temporally limited strengthening of their own fighting power. Often they took on a planned character, were organized for long terms by offices at various levels, and were meant to compensate for existing shortages of armaments and equipment.

The German troops captured more than 100 British tanks at Cambrai from November 20 to 28, 1917. Here a Mark IV (female) is seen in German service in February 1918.

Captured Weapons from Spain, Austria, Czechoslovakia, and Poland

In the German Wehrmacht, the reasons for the reuse of captured war materials were similar to those of World War I. Between 1918 and 1936 there was no access to them, other than the equipment that had been obtained for testing by the Army Weapons Office, sometimes in very roundabout ways. It is interesting that, in the specialist press and literature, there were frequent references to the use of captured weapons by German troops in World War I. What was not to be found were general experiences and conclusions. In 1937, D 98, "Foreign Weapons", was published. This was merely a catalog of types which, based on foreign press releases and divided into various subjects, was meant to make identification of foreign weapons easier.

From 1936 on, infantry weapons, artillery guns and armored weapons that had been captured by German troops in combat were once again available. They came from Spain, where the Condor Legion fought on the Nationalist side against the forces of the People's Front fovernment. The center of interest was the modern war materials that had

been delivered to the Spanish Republic by Soviet Russia. Its evaluation was based on two main points:

1. Examination of the technical experiences outside Germany by the Army Weapons Office and the Industry. Parallel to this, information was to be gained as to the effectiveness of foreign fighting techniques and the possibility of opposing them.

2. The reuse and maintenance of captured weapons and vehicles within the Condor Legion. This applied above all to the light T-26 tank, which could not be matched by the German Panzerkampfwagen I (Sd.Kfz. 101).

After the union of Austria with Germany in the spring of 1938, the Wehrmacht took over the weapons and equipment of the Austrian Army. This was a welcome gain, as Austria had obtained licenses to build several modern weapons, and had built them along with noteworthy Austrian-developed types. They [p. 9] included pistols (Steyr), rifles and carbines (Mannlicher), and machine guns (Schwarzlose).

The Spanish Nationalists, in the course of the bloody Civil War on the Iberian Peninsula from 1936 to 1939, captured a great number of usable Russian T-26 tanks. Some of them were inspected by the German Army Weapons Office.

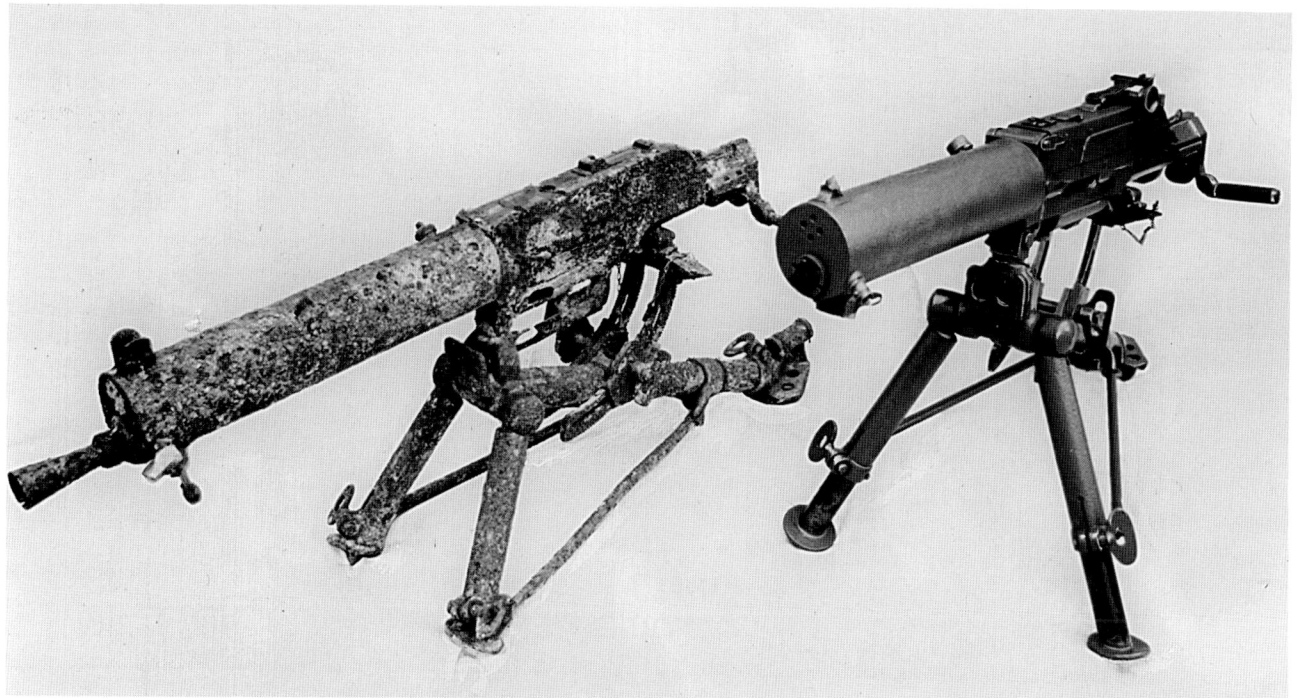

Schwarzlose machine guns from the armies of various European countries came into German hands. At left is a 247(j) machine gun, a Yugoslav Schwarzlose 7/12, at right an 07/12 (ö) machine gun without a flame retarder.

The Austrian artillery, for the most part, remained from World War I, but had been reworked in the thirties by firms like Böhler in Kapfenberg, and was thus in good condition. The guns were put to use by the Wehrmacht. A few were still being used by the artillery replacement units at the end of World War II.

Before the Anschluss, Austria had bought eight 4-cm anti-aircraft guns from the Swedish firm of Bofors and obtained a license to produce another fifty guns with 235 replacement barrels. In 1939 the German Luftwaffe requested that the production of these high-performance weapons be continued at the Vienna Arsenal at the rate of 20 per month.

Also taken over were twelve Oerlikon 2-cm anti-aircraft guns with 36,000 rounds of ammunition, which Austria had bought in Switzerland.

One of the first foreign trucks used in large numbers by the Wehrmacht was the light, off-road-capable Steyr 640 1.5-ton (5x4) truck. In all, 3780 of them had been built. This vehicle was most often found with the 2nd and 9th Panzer Divisions (in Vienna and St. Pölten). Also of significance was the Austro-Daimler ADGZ (8x8) heavy armored scout car, which also saw service as an armored police vehicle with the SS-Ordnungspolizei. A large number of the Saurer RR 7/2 halftracked armored vehicles which were built in Austria were used by the Army as armored observation vehicles for the artillery. Tracked vehicles were also also taken over and used by the mountain troops.

On March 15, 1939, Wehrmacht units marched [page 11] into Czechoslovakia. The Czech war materials were of excellent quality and was of great importance in the expansion of the German Army before World War II. Taken over by the Wehrmacht were, among others:

Captured Schwarzlose machine guns, on makeshift twin mounts, were used by this Luftwaffe unit in the east to secure airfields against low-flying air attacks. The picture dates from the winter of 1941-42.

Two-centimeter anti-aircraft guns made by the Swiss firm of Oerlikon came not only from Austrian war materials but also from the armaments of other armies. This picture was taken in a German support point on the Atlantic coast in 1943.

From Czechoslovakia, Germany took over more than 500 anti-aircraft guns, including a large number of 8.35 cm 22 (t) guns. Because of their insufficient aiming gear, they were used during the war by barrage-fire Flak batteries in the German war zone.

480,000	Pistols, type 27
31,204	Machine guns, type 26
514	8 cm grenade launchers, type 36, with 236,500 grenades
1123	3.7 cm antitank guns, type 37 (still made until May 1940)
273	4.7 cm antitank guns, type 36
184	6 cm cannons, type 30, with 240,000 rounds
47	10 cm mountain howitzers, type 16/19, some later rebuilt to take German ammunition
108	10.5 cm cannons, type 35, with 68,600 rounds
127	10 cm field howitzers, type 30
42	15 cm heavy field howitzers, type 15, with 20,300 rounds
219	15 cm heavy field howitzers, type 25, with 98,000 rounds
238	15 cm heavy field howitzers, type 37
22	21 cm mortars, type 18/19, with 6900 rounds
10	24 cm howitzers, type 39
200	anti-aircraft machine guns, type 39
218	LT-35 tanks
9	LT-38 tanks (end of May 1939 specifications; series production of this model had just begun at that time)

Among the Czech booty there were large numbers of motor vehicles, plus field kitchens, field bakeries, aircraft, and weapons from fortresses. The 3.7 and 4.7 cm anttank guns in particular, as well as tank turrets of various designs, were of particular importance for the construction of the defensive fortifications on the German-French border, the West Wall, which was being pushed at that time. Some of the cannons, similarly to the Austrian Böhler guns, were mounted on makeshift turntables of wood in antitank positions; others were set on steel mounts. On July 7, 1939, in the area supervised by the engineer officer for the western German fortifications, there were 422 antitank positions in all, of which 145 145 used the makeshift turntable mounts mentioned above. The Army High Command had prepared, at least in terms of organizing, to establish nine infantry divisions of the 5th and 6th waves in the autumn of 1939, for the expansion of the war in the west. As is known, France and Britain entered the war against Germany on the Polish side. The establishment of these nine divisions was urgently needed. They were given mainly Czech weapons and equipment. Also gaining from the wealth of captured materials were several Panzer and artillery divisions then being established. In addition, Himmler's SS took over a considerable part of the Czech war materials and could thus hasten the expansion of its front and police units without depending on the Army's supply lines.

In Czechoslovakia, 3158 of the 10 cm light Field Howitzer 30 (t) had been built. Of them, 127 were in use by SS units in 1940. A temporary description of them (D 370) was issued on July 11, 1941.

In order to differentiate the Czech weapons and equipment from the German material, which sometimes had the same model numbers, they were given the suffix (t) (= tschechisch), a practice later extended to captured war materials from other countries.

The camppaign against Poland in September 1939 was only of short duration and caused the Army comparatively few weapon and equipment losses. Here too, considerable quantities of war materials were captured. Alomst all the Polish handguns, as well as the machine guns, were made for ammunition available in Germany, which made their re-use easier. Among the captured heavy weapons were:

621	3.7 cm antitank guns, type 36 (of which 556 were sold to Romania)
721	7.5 cm cannons, type 97 (80 were sold to Romania)
676	10 cm howitzers, type 14/19, with 471,000 rounds
171	10.5 cm cannons, types 13 and 29
14	22 cm mortars, type 32

Polish armored and other motor vehicles were of little importance. Some of them were used by the occupation troops. The other arms of the Wehrmacht, the Luftwaffe and the Navy, took the trouble to examine the captured material, sort it out by genres, obtain and reuse it. Insight into this laborious work is provided by a report by the Luftzeugmeister Ost, according to which there were assembled between November 1 and 10, 1939:

26	Aircraft (damaged)
1400	Aircraft tires
55	Aircraft engines
37	4 cm guns
3	7.5 cm guns
51,165	4 cm explosive grenades
86,300	1 kg bombs
10,250	12 kg bombs
4660	100 kg bombs
1160	[Fliegerpfeile] etc.

Soldiers cleaning weapons, photographed in the spring of 1940. The handguns and machine guns are of Polish and Czech origin.

The 15 cm heavy Field Howitzer 25 (t), of Czech origin, was used in 1939-40 to arm new units of the artillery regiments in several infantry divisions. This picture was taken on the upper Rhine in the winter of 1939-40.

The 15 cm heavy Field Howitzer 15 (ö) or (t) was found much more rarely in the German Wehrmacht. It came from Austrian and Czech army stocks. The picture shows such a gun in action on the eastern front in the winter of 1941-42.

A 7.5 cm Anti-aircraft Gun 37 (t) in firing position. It was originally planned to turn this modern gun over to Yugoslavia.

Six heavy 24 cm cannons also came from the Czech Army; their gain was already arranged by the Reich Foreign Ministry in December 1938. The guns saw action only with the II. Unit of Heavy Artillery Regiment 84. This picture was taken in Belgium in May 1940 (compare WA No. 138).

The 30.5 cm mortar (t) could fire 289-kilogram shells to a distance of 12,300 meters, and was especially good for attacking fortresses. 21 guns were used in June 1942 in Operation "Störfang" (the capture of the Sevastopol fortifications) in Crimea.

Below: Parts of Panzer Regiment 11 on the eastern front in June 1941. At that time, 187 formerly Czech LT-35 tanks, designated 35 (t), were in service with the Army.

Left: Using Czech antitank guns, 132 tank destroyers were built in 1940 on the Panzer I chassis. The official designation was "4.7 cm Pak (t) auf Panzerkampfwagen I". The Czech gun, with a penetrating power of 47 mm of armor plate at 500 meters, exceeded the power of the German 3.7 cm antitank gun (29 mm at 500 meters) considerably.

Below: From the Czech armor builders, the Army obtained 1073 Panzer 38 (t) tanks, of which 665 were put into service with five Panzer divisions on June 22, 1941. In the ensuing years, the robust chassis were used for self-propelled gun mounts; 600 turrets reached Army armories. Of them, 361 were still on hand in March 1945. The majority of these turrets were used for special operations in fortresses and defense areas. The turret shown here belonged to Armored Turret Company 1312 in Frankfurt on the Oder in April 1945. On August 24, 1984 it was removed, and it is now in a display of the military History Museum of Dresden at the fortress of Königstein.

Above: A Czech Avia-Fokker F IX bomber in German service.

Left: Most of the 37 mm Type 36 antitank guns captured in Poland were sold to Romania. Only 65 guns, designated 3.7 cm Panzerabwehrkanone 36 (p), were used by the Wehrmacht. This one was used by a Luftwaffe unit.

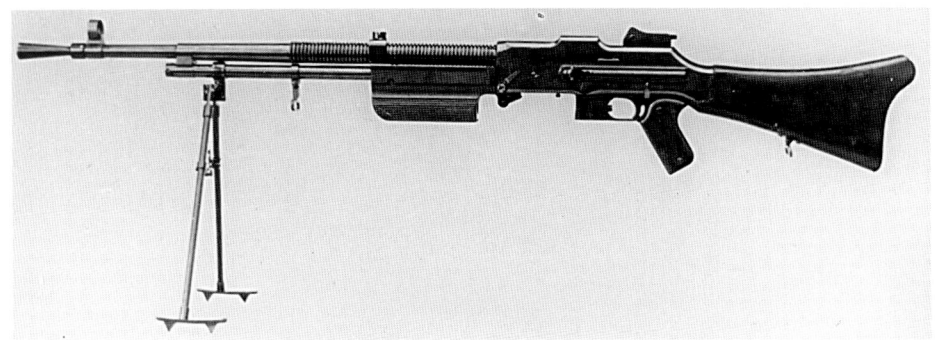

Left: With the Browning 28 Light Machine Gun 154/2 (p) (Polish), one could fire 7.92 mm ammunition made in Poland or Germany. The weapon weighed 9 kilograms, and its rate of fire was 500 rounds per minute (in theory).

Above: The Type 30 Polish heavy machine gun first came into German hands from supplies of the Polish Army, then, after June 22, 1941, from Soviet Army stocks. Its designation was "schweres Maschinengewehr 249 (p)-poln. Browning 30-".

Limited numbers of the 4.6 cm Grenade Launcher 36 (p)-poln. No. 36- were used by the German Army. This launcher weighed 12.6 kilograms and had a range of 800 meters.

Left:The Polish 220 mm Type 32 mortar had been built by the Czech firm of Skoda. The Wehrmacht took over 14 guns with 1825 shells, designating them 22 cm-Mörser 32 (p)-poln. wz 32.

Right: This Polish ammunition cart, made to transport grenade-launcher ammunition, was photographed in 1939 by order of the the Army Weapons Office, which received the captured field vehicles and studied their suitability for use.

Captured Weapons and Equipment from Denmark and Norway

The Wehrmacht's operations in Denmark and Norway in the spring of 1940 brought in captured war materials of various quality. Particularly worthy of note are the 2 cm anti-aircraft guns made by the Danish firm of Madsen, 189 of which were still on hand on February 11, 1943. The gun was used mainly by the Navy. In 1942, the M.Dv. No. 170.2 "Merkbuch über die Munition für die 2 cm Flak Madsen" was issued. Production of the ammunition was continued.

In 1943-44 the Army obviously had to make use of still-stored remaining stocks of captured ammunition. At the front in Italy, the Danish 1923 model hand grenade made its appearance at that time. Early in 1943 there were still 47,623 of them on hand. In addition, handguns and machine guns were being used widely by the Wehrmacht. It is noteworthy that the 1903/24 model light machine gun—later given the German number 158 (d)—that had been used by the German Army already in World War I.

The artillery guns captured in Denmark were, like those from Norway, rather few in number. Among the guns that came from Norway, the 7.5 cm Field Cannon 01 gained particular significance, as did two models of 12 cm howitzers.

The long-range cannons of the Norwegian coastal batteries, with a caliber of 15 cm or more, were sometimes mounted in screen mounts, and were used by the naval artillerymen and for coastal defense on the Atlantic Wall.

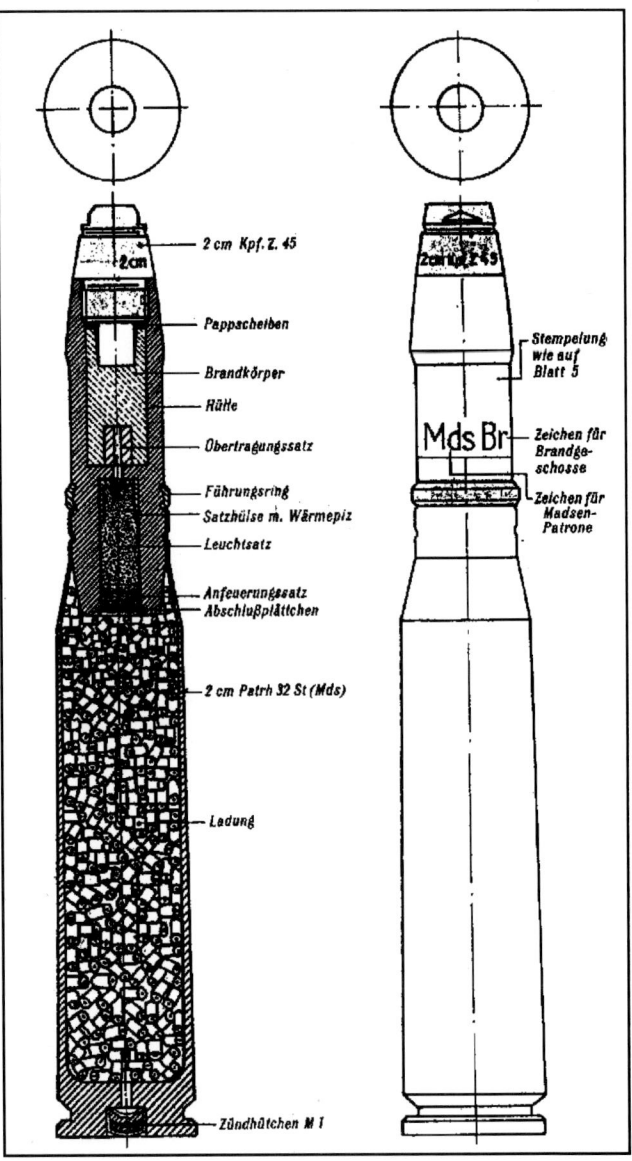

Drawing of an incendiary shell with tracer light for the 2 cm anti-aircraft guns of the Danish firm of Madsen.

Captured hand grenades that were used by the Wehrmacht. From left to right: the Danish 342 (d)-dän- 23, an Italian, and a French grenade.

Weapons and Equipment Captured in Belgium, the Netherlands, France, and Britain

On May 10, 1940, the Wehrmacht began its attack in the west. The fighting forces of Belgium, the Netherlands and France were thoroughly beaten in a matter of weeks. The British Expeditionary Corps was able to escape across the Channel, leaving almost all its heavy weapons, vehicles and equipment behind. The quantities of captured war materials exceeded everything taken previously. In an inclusive report of the Wehrmacht High Command about the operations in the west from May 10 to June 4, 1940, it was stated: "The weapons and equipment of some 75 to 80 divisions, with guns up to the largest calibers, tanks and other vehicles of all kinds, were destroyed or captured." By June 25, 1940, besides the previous booty, all the arms and equipment of some 55 further French divisions fell into German hands. To this, the materials of the Belgian and Netherlands troops and the British expeditionary corps must be added.

The quantity of handguns was significant. From the very beginning, problems arose in supplying ammunition for the calibers of the Belgian, Netherlands, French and British handguns and machine guns, which differed from the German calibers. Among others, these included 7 mm, 7.5 mm, 7.65 mm, 7.71 mm and 8 mm calibers, as well as 12.7 mm and 13.2 mm for the heaviest machine guns. At first the numerous ammunition stores made the situation appear in a different light. This was to change as the war went on. It is interesting to note the use of the British STEN machine pistol. At first the weapon had been rejected by the

German forces on account of insufficient safety when in use. In 1944 the Reich Safety Office ordered the distribution of large numbers from storage, some to the high SS and police command in Greece, so they could use them in action against partisan bands. At the end of 1944, further construction of these weapons began. In particular, the Volkssturm was to be armed with them.

The numerous Belgian and French pistols helped to fill the need for handguns in the three branches of the Wehrmacht.

The French artillery, which had the reputation of being a technically and tactically outstanding weapon in World War I, was undoubtedly still very strong in 1940. But obsolete guns with insufficient ranges prevailed. In addition, the French often used bag cartridges for propellant charges. In the Wehrmacht, capsule cartridges were preferred, for safety reasons. Essentially, this evaluation also applied to the British guns. Even so, the numerous captured artillery weapons were put to many uses by the Wehrmacht.

The Army Weapons Office took over 823 of the 47 mm 37 SA antitank gun. In 1940 they were among the most effective antitank weapons and strengthened the tank-destroyer units of numerous infantry divisions of the Army. In the Atlantic Wall, this weapon often was used in fixed positions in Regulation Building 139, in loophole positions.

Many of the 75 mm 1879 field cannon were on hand. Without counting those from Polish and Belgian stocks, the

Some vehicles, machines and equipment from civilian ownership were taken over by the German Wehrmacht, including this truck of a Belgian entrepreneur, photographed in 1940.

Several DAF armored scout cars from Holland were put to use by German reconnaissance units with the designation DAF 201 (h). This vehicle was lost on the Eastern Front in the winter of 1941-1942.

Wehrmacht still had over 683 weapons of this type in March 1944.

In 1942 the Army Weapons Office already had ordered 860 1897 field cannons rebuilt into 97/38 tank-destroyer guns. Their barrels were set in mounts of the 38 and 40 antitank guns.

Among the other captured guns, there were some 1700 155 mm howitzers, 200 220 mm type 16 TR mortars, and about 700 155 mm cannons of various types. Many were put to use in the coastal defenses of the Atlantic Wall. Others, like the 220 mm type 16 TR mortar, were used by units of the army artillery. As can be seen in a memo of July 8, 1943 from the General Command, XXXVIII Army Corps in the east, sixteen 22 cm 531 (f)-frz. 16 TR mortars were used to form artillery focal points. The crews had to be formed from local artillery units.

The proportion of captured cannons in the army coast artillery was particularly high. The staff of Artillery Regiment 781 in the west had eleven units with the following equipment:

10 batteries with 10.5 cm 35 (t) cannons
2 batteries with 15 cm 15/16 (t) cannons
13 batteries with 15.5 cm 418 (f) cannons
6 batteries with 15.5 cm 416 (f) cannons
8 batteries with 15.5 cm 420 (f) heavy field howitzers

French artillery guns dominated. British guns were of much importance only for a time in North Africa, where the artillery of the 90th Light Division was composed of 8.76 cm Type 280 (e)-engl. 25 pdr MK IV and MK V field cannons.

Many guns, other than anti-aircraft guns, were scrapped. The use of the captured Belgian guns, which went up to the 28 cm Cannon (E), and a good many of which were of German origin, had a different fate, as they could continue to be used without any problems in obtaining ammunition.

In May 1940, Railroad Battery 655 had been equipped with two of the 15 cm Cannon (E). Both guns were lost to barrel damage after just a short time. Then the battery was issued two of the 28 cm Cannon (E) of the Belgian Army; they had been captured at Tirlemont and Hasselt, along with 133 shells. These were German railroad cannons from World War I, which meant that they could be put into service very quickly with the help of the Krupp firm. As of May 30, 1940 they saw service in the battery of Army Group B (above). On June 16, 1940 one gun lost its barrel to an explosion (below). The barrel landed near the rails 42 meters ahead of its mount.

This tank turret of French manufacture, photographed in July 1940, was part of the Belgian defenses on the coast near Ostende. It later became part of the Atlantic Wall.

Left: The 22 cm Mortar 530 (b)-belg. TR 16 S is identical to the French 220 mm Mortar TR mle. 1916 Schneider. Both were used by units of the army and fortress artillery, and could fire 100.5-kilo-gram shells up to 10,800 meters.

On May 10, 1940 the French fighting forces had 3132 tanks. A large number of these needed repairs but were captured in usable condition. According to plans of November 5, 1942, it was intended that by the spring of 1942 the following French tanks would be ready for service:

500 Renault FT 17/18
60 Renault R-35, rearmed with 4.7 cm Panzerabwehr-kanone (t)
125 Renault R-35
200 Hotchkiss
20 Somua

In addition, 400 Hotchkiss and 20 Somua tanks were planned for Captured Tank Brigades 100 and 101. But the inclusion of French tanks and their use according to German principles caused problems. It is stated in the aforementioned plans: "In order to evaluate usefully the great number of captured tanks, the use of which in German tank units was rejected because of their characteristics, which are very inferior to the German tanks, it was decided that they should be used for the securing of occupied areas. . . " But there were also other reasons that resulted in the captured French tanks only being used further under certain conditions. They included the extreme loudness when firing of the mostly cast tank bodies. In the Army Weapons Office's firing tests in the summer of 1940, up to 127 phones were measured inside the tanks, representing the limit of physical endurance.

Thus a considerable number of the French tanks were sent to military repair shops or to the industry for special uses. These included tank destroyers, gun vehicles and grenade-launcher carriers. The progressive deterioration of the Army's motor vehicle situation in the autumn of 1941 led in November to 100 Renault and 60 Somua tanks being rebuilt into makeshift towing vehicles. A similar special action in the summer of 1942 also included British and Belgian vehicles.

In August 1940, Hitler already decided that in the further enlargement of the Army, the possibility of a campaign against Soviet Russia had to be considered. By the time this campaign began in June of 1941, 84 more divisions were created. Without the extensive booty from the western campaign of 1940, these units would have remained without weapons and vehicles. Motor vehicles in particular played an important role in the motorization of at least some of the divisions, even though they [p. 27] were not to have long lives under the rough conditions of the eastern theater of war. The Panzer divisions suffered particularly. The 18th Panzer Division, which was equipped with strictly stock French motor vehicles until the end of May 1941, could assure its mobility only with difficultyn and by using its en-

As in World War I, captured French machine guns were used for anti-aircraft defense. The weapon was designated 8 mm heavy Machine Gun 257 (f)-frz. Hotchkiss 14-. The model also came into German hands from Polish, Yugoslav and Greek stocks.

A collection point for captured French war materials, photographed in July 1940. Of the 47 mm Antitank Gun mle. 37 SA shown here, the Army Weapons Office took over 823 in all. The guns, weighing 1070 kg, exceeded the German 3.7 cm Antitank Gun L/45 in terms of penetrating power against armor plate by 21 mm (= 50 mm at 500 meters).

This 4.7 cm Antitank Gun 181 (f)-frz. 37 SA-APX- was used by an infantry tank-destroyer company which was located in the vicinity of Warsaw in the spring of 1941. On February 28, 1941, the D 337 "Preliminary Device Description and Service Manual" for this gun was issued.

The captured Antitank Gun 181 (f)-frz. 37 SA-APX- was reequipped according to German principles of use. This included a makeshift third wheel. In 1941 the 4.7 cm Panzer Shell 40 was also introduced for this gun. Thus its penetrating power was increased to 70 mm. With the 4.7 cm Stick Grenade 42, 180 mm could even be attained, though only at a range of 100 meters.

tire repair-service capacity. Among the trucks, the 4.5-ton Citroen Type 45 attained a certain significance.

The one-ton Peugeot and four-ton Matford were also seen often. The same was true of the French halftrack towing vehicles, which were used as tractors in the Panzerjäger units, infantry gun companies and motorized artillery units. The Czech emergency service also received such vehicles.

Captured British vehicles were especially popular and regarded as very robust. Great parts of the German Afrikakorps supplied their fighting forces with vehicles captured from British Commonwealth stocks.

Captured French aircraft and on-board weapons were put to use within certain limits. The Luftwaffe made more frequent use of airdropped ammunition. Among others, the French 50-kilogram splinter bombs in packages of four with the Ab 500 3 A airdrop container were used, and the small 1-kg splinter bombs were also kept in production.

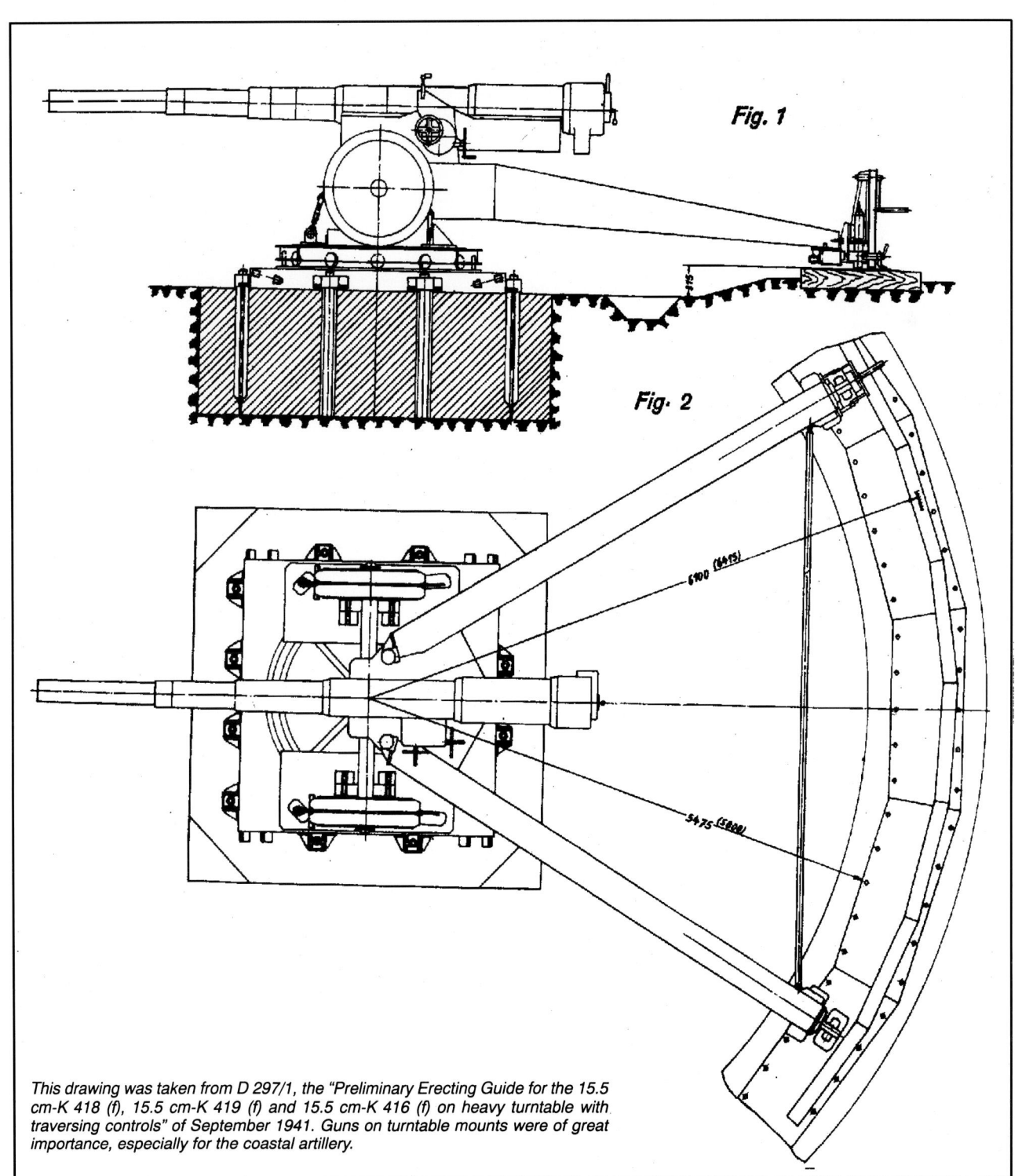

This drawing was taken from D 297/1, the "Preliminary Erecting Guide for the 15.5 cm-K 418 (f), 15.5 cm-K 419 (f) and 15.5 cm-K 416 (f) on heavy turntable with traversing controls" of September 1941. Guns on turntable mounts were of great importance, especially for the coastal artillery.

Left: The 15.5 cm Cannon 416 (f)-frz. L 17 S- in a firing position on the Atlantic Wall, photographed in 1943. The gun could fire heavy shells up to 17,300 meters. It weighed 8695 kg ready to fire.

For the 15.5 cm Cannon 416 (f)-frz. L 17 S-, bag cartridges, kept in containers such as this, were used. There were both small and large charges, and these came in both day and night charges. The day charge consisted of 8.72 kg of powder, to which three additions of 0.5 kg each were added for night use.

A battery of 15.5 cm Cannon 416 (f)-frz. L 17 S- in firing position on the eastern front. Such batteries belonged to the Army artillery and were partially mobile. The guns were towed by German 12-ton towing tractors, or sometimes by captured towing vehicles.

A 19.4 cm Cannon 485 (f)-frz. GPF- in firing position on a self-propelled mount. The generator vehicle has been uncoupled and the gun car lowered. One gunner is handing a bag cartridge up to the work stage. The picture was taken in the summer of 1942 in the II. Unit, Army Artillery Regiment 84, in the northern sector of the eastern front.

Although it was obsolete, the German Wehrmacht also adopted the 16.4 cm Cannon (E) 453(f)-frz. 93/96 M-, which could fire 49.8-kg. shells up to 19,200 meters.

In May 1940, the French Army had 476 Schneider Creusot 220-mm Mortar TR mle. 1916 on hand. A goodly number of them were taken over by the German Wehrmacht and designated 22 cm Mortar 531 (f)-frz. TR 16 S-. Some of them were still usable at the end of World War II. The 4th Battery of Fortress Artillery Unit 1325 in Frankfurt on the Oder had four mortars of this type on April 20, 1945. Three days later, the last mortar was blown up in the vicinity of Märkisch-Buchholz before the 9th Army broke through to the west.

The mantle and the breech were found in 1983 and can now be seen at the Military History Museum of Dresden's display at Königstein Castle.

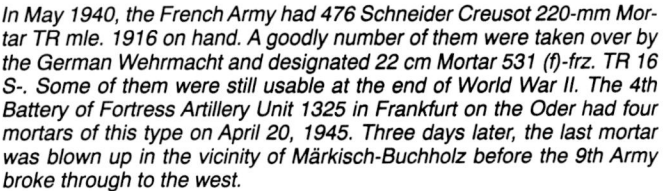

This shell for the 52 cm Howitzer (E) 871 (f)-frz. 16-, found in the state of Brandenburg in 1994, weighs 1654 kilograms. Captured guns of this type were already being used by the Army Group Morth in the winter of 1941-42.

The 7.5 cm Antitank Gun 97/38 shows that antitank troops needed to obtain better weapons as fast as possible in 1942. The barrels of numerous captured French mle. 1897 Cannons were installed in the mounts of the 5 cm Antitank Cannon 38. At first, ammunition captured from Poland and France was used. With the German-made 7.5 cm 97/38 HL B (f) shell, armor 75 mm thick could be penetrated. In all, 860 97/38 Antitank Cannons were made.

The Simca Cinq car was taken over from French military stocks. The car, with a 13 HP engine, was not up to the rough conditions of front-line use. It lasted longer with training and replacement units.

Right: One could see the Citroen Type 23 (4x2) truck more often in the east too. The French Army had over 12,500 trucks of this kind, and a goodly number of them came into the hands of the German Wehrmacht.

The Renault BDS-2 was used as a communications and staff car by the German occupation troops in France. The four-cylinder engine produced 52 HP and could attain 125 kph.

The DEP Diesel-electric snow tractor was made in France. In February 1940, the French Army tested such vehicles of various brands in the Alps. Their further development resulted in vehicles like the DEP model, which were built with Deutz Diesel engines to the order of the Inspector-General of German Highways. This picture was taken in Norway in 1943.

Left: A Somua S 303 (f)frz. tractor towing a light 3-ton tractor (Sd.Kfz. 11) on the eastern front in the spring of 1942. The French Army had 308 of these Somua MCL 5 tractors, used for towing purposes, in service on May 11, 1940. In German service they were utilized by the Army and Technical Emergency Service.

The Unic P 107 304 (f)-frz. towing tractor was used particularly to tow antitank cannons and infantry guns. The picture, taken on the eastern front in July 1941, shows such a towing vehicle towing a 3.7 cm L/45 Panzerjäger cannon.

In the return from France, this Panhard 38 P 204 (f)-frz. furnished as a command vehicle (without a turret) was taken along. When these vehicles were not listed in war-strength records, they were generally sent off to collection places for captured materials.

A Panhard 38 P 204 (f)-frz. armored scout car is being towed on the eastern front in the summer of 1941. The Panhard armored scout car was a modern vehicle, of which more than 100 came into German hands. They were used by reconnaissance units, as railcars in conjunction with armored railroad trains, and by the police.

The UE 630 (f)-frz. Chenillette- armored ammunition carrier was used primarily by the infantry divisions of the Army, in order to make heavy infantry weapons mobile. Here a 3.7 cm Panzerjäger Cannon L/45 has been mounted on the towing vehicle. Such makeshift solutions were practiced mainly by advance units, to achieve greater mobility and readiness to fire. The picture was taken on the eastern front in the summer of 1941.

Right: This UE 630 (f) armored ammunition carrier tows a 5 cm Panzerjäger Cannon 38 for an Infantry Panzerjäger company on the eastern front, in the summer of 1941.

This Hotchkiss H 38 Driving School Tank 38 H 735 (f)-frz. was used by a replacement troop unit in Romania in the spring of 1944.

This makeshift armored train, used for securing in the Balkans in 1941, was equipped with a Renault 18 R 730 (f) tank, which could also have been captured in Yugoslavia.

Between May and October 1941, the Alkett firm mounted 174 4.7 cm Pak (t) guns on R 35 (f) tanks as tank destroyers and rebuilt another 26 as command cars. The vehicles were used by, among others, Army Panzerjäger Unit 561. This picture shows a command car with MG 34 in a Ball Mantlet 30, followed by two tank destroyers with 4.7 cm guns.

The "Delay Pencil" long fuses were dropped over the occupied areas in the west by the British for sabotage purposes, and thus came into German possession.

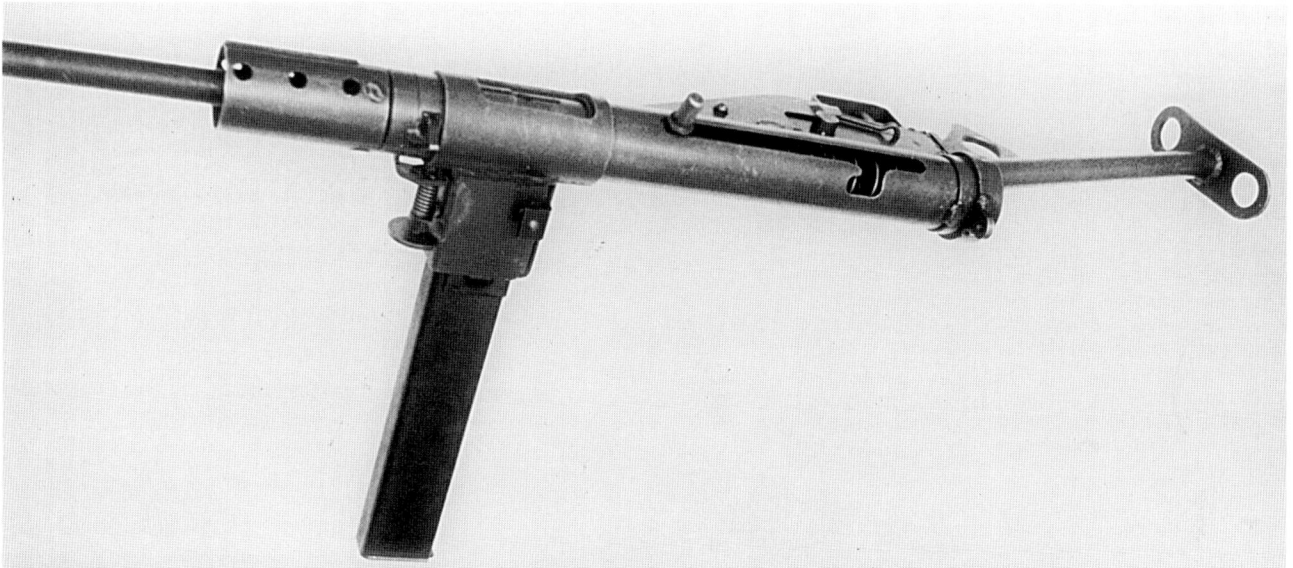

The STEN Mk. II Machine Pistol 749 (e)-engl. was captured in large numbers by the Germans, along with sabotage weapons. This gun could also fire German-made 9 mm ammunition, and was rebuilt and turned over to the Volkssturm toward the end of the war.

A light 1-ton Towing Tractor (Sd.Kfz. 10) was used to tow this 8.76 cm Field Cannon 281 (e)-engl. 25 pdr Rohr Mk 1, Laf. 25/18 pdr. Mk IV (in North Africa in 1942.

A weapon that came from British World War I supplies was the 4 cm Vickers anti-aircraft gun of 1918. The rate of fire was about 100 rounds per minute. This picture was taken on the French Atlantic coast in 1943.

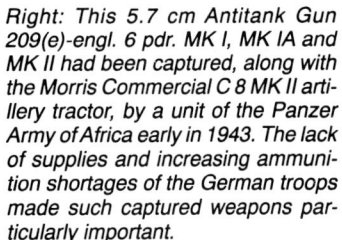

Right: This 5.7 cm Antitank Gun 209(e)-engl. 6 pdr. MK I, MK IA and MK II had been captured, along with the Morris Commercial C 8 MK II artillery tractor, by a unit of the Panzer Army of Africa early in 1943. The lack of supplies and increasing ammunition shortages of the German troops made such captured weapons particularly important.

This 5.7 cm Antitank Gun 209 (e)-engl. 6 pdr. MK III- was captured from the British in Normandy early in July 1944. A Lloyd Carrier is towing it.

Robust and reliable: a British 15 cwt Morris Commercial CS-8 (4x2) truck used as a German unit's command vehicle on the eastern front in July 1941.

These Morris Commercial C 8 MK II artillery tractors had to be left behind by the British when they evacuated Greece. After a thorough overhaul and camouflage paint, they were given to a light howitzer unit, horsedrawn until then, as towing tractors.

In the spring of 1942, the Army tried particularly to make the Panzerjäger units deployed on the southern part of the eastern front more mobile by issuing them towing vehicles. This picture, taken in the 14th Panzer Division's sector in June 1942, shows a MK II 603 (3) light artillery tractor towing a 7.5 cm Antitank Gun 40 L/46.

The Br 732 (e)-engl. Bren Carrier was used in the Wehrmacht as a small armored machine-gun carrier.

The Weapons and Equipment Captured in Yugoslavia and Greece

In the course of the offensive operations in the Balkans in the spring of 1941, the Wehrmacht captured important parts of the armaments and equipment of the Yugoslavian and Greek fighting forces. Much of it, as could be seen in a display of just the materials captured on the Lamia Pass road to Larissa by the Leibstandarte Division, was British material.

Along with handguns and machine guns, there were artillery materials in particular, made in Czechoslovakia and France, among the Yugoslav and Greek booty. It was no problem to use them along with materials captured from those countries and already on hand.

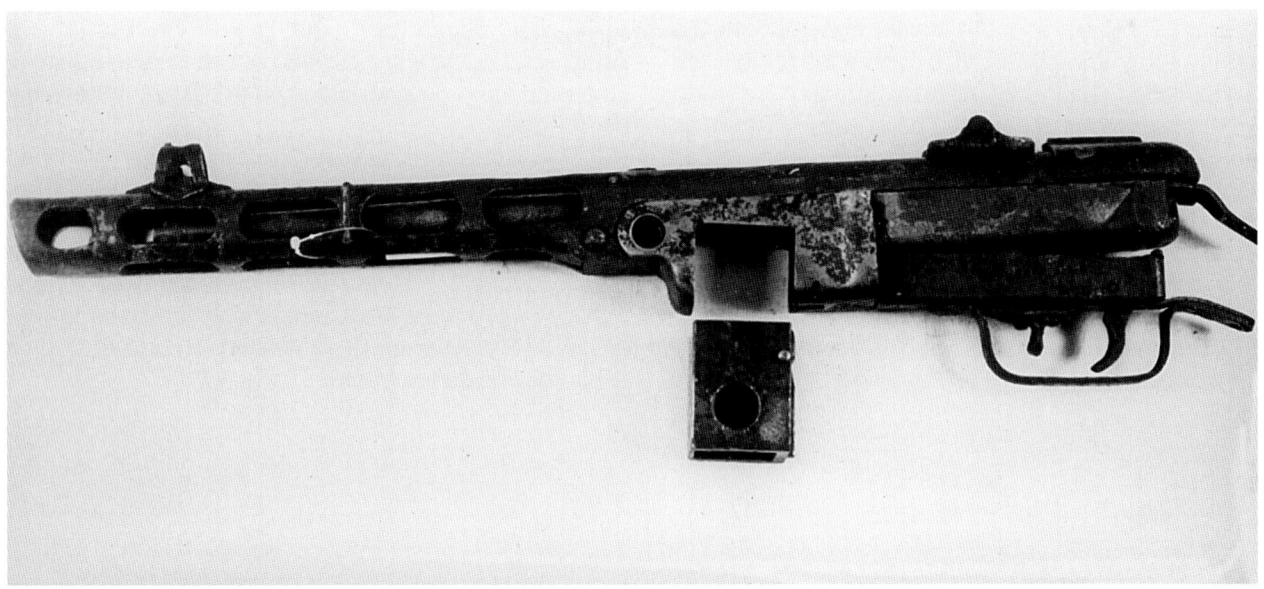

The excellent Russian machine pistol was used by the German Wehrmacht under the designation of Machine Pistol 717(r)-russ. M 410. The fragment shown here bears a Waffen-SS stamp on its magazine holder and was set up to take German 9 mm ammunition.

The Weapons and Equipment Captured in Soviet Russia

The previous experience with already captured or expected foreign equipment had been evaluated by the Army Weapons office. To be able to supply the troops with usable supplies, the Army High Command, beginning in March 1941, issued the D 50 "Manuals for Foreign Equipment". They were divided into numerous genres and were constantly added to. All the "foreign-land" war materials, other than those of Austrian and Czech origin, were assigned uniform designation numbers (captured-equipment numbers). As far as is known, the foreign materials expected to be captured in the campaign against Soviet Russia were also included.

The war against Societ Russia began on June 22, 1941. If one believes the Wehrmacht report, within two months a gigantic arsenal of Russian was materials had been captured or destroyed, including 14,000 armored vehicles and 15,000 guns. Some of the guns, grenade launchers, tanks, etc., were of excellent quality. What was published in the periodical "Deutsche Wehr" in 1942 was quite true: "The guns possess all the marks of modern construction, . . ."— all the more reason for the collecting places to collect Russian materials eagerly. But there were problems involved. The tanks captured by the thousands in the east were supposed to be put into German service, as the French tanks

had been. But often their use was ruled out by their great weight, which went up to 52 tons, making them too heavy for German towing vehicles. The available repair services scarcely had the capacity to handle German tanks, and spare parts were also lacking. Thus by the end of October 1941, only slightly more than a hundred Russian tanks were being used for securing tasks.

Because of their great numbers and good quality, the captured Russian guns took on an especially great importance for the Wehrmacht. They included field guns, light and heavy howitzers, cannon howitzers and cannons. The Army had at least 974 of the 15.2 cm Cannon Howitzer 433/1 (r)-russ. 37. In February 1943, production of ammunition for them began. The same is true of the 12.2 cm Heavy Field Howitzer 396 (r)-russ. -38, ammunition for which was made as of April 1943. The 12.2 cm Cannon 390 (r)-russ. 31/37 also filled [p. 42] needs for German artillery weapons, especially for coastal defense. There were 424 of them on hand.

In the spring of 1942, several artillery units were equipped with Russian 7.62 cm field cannons to compensate for the losses in the winter's fighting. Their great range and effective use against tanks were counteracted by an insufficiently variable shell trajectory and the limited possibility of observing their firing and thus making quick changes. Very similar judgments were made in the Panzer Instructional Division, which used 15.2 cm cannon howitzers in their artillery regiment in France in 1944.

Reports came repeatedly from the troops, stressing the superior range and effect of the Russian company, battalion and regiment grenade launchers. Their reuse by the Germans, though, was urgently advocated at times because of the shortage of ammunition for the German launchers.

The Russian 120 mm Regimental Grenade Launcher 38, designated 378 (r), was put to use, and performed so well that the Army Weapons Office contracted for its copying (12 cm Grenade Launcher 42) in September 1941. Many other weapons were also copied. These included the FOG fixed-position flamethrower (Defensive Flamethrower 42) and the wooden-boxed rifle mine (Rifle Mine 42), to name just a few examples.

Between early 1942 and early 1944, 490 of the 7.62 mm Field Cannon 36, captured in great numbers, were reworked so their chambers could take the shell of the 7.5 cm Antitank Cannon 40. The thus-created antitank weapon was used on wheels or self-propelled mounts. The Chief of the Army General Staff said of them on August 5, 1942: "The Russian 7.62 cm field gun has led to the best success for the Germans after being reworked as an antitank gun . . ."

Of the smaller Russian weapons, the self-loading rifles and the PPSch 41 machine pistol were especially popular, the latter because of its ruggedness and its large-capacity cartridge drum. Existing documents show that the SS Weapons Office made efforts to adapt this weapon for the 9 mm Parabellum shell.

Some of the Russian guns were put to use by Army units, while others were used to arm the Russian Vlasov Army , and some were turned over to local securing units. Captured hand grenades, mines and other materials were used widely.

The Russian military vehicles were used gladly by the German troops despite their poor performance. This was because of their resistance to the terrain and weather conditions in the eastern war zone, which also applied to the artillery tractors, which were sometimes used for special tasks. For example, the Koenigsberg Fire Department was assigned the Russian Light Artillery Tractor 601 (r) -russ. ST 3- for use as fire trucks.

The Russian Heavy Machine Gun M. 1910 was also captured in quadruple anti-aircraft gun form. It was used as number 216 (r)-russ. M 1910.

Left: With a penetrating power of 30 mm of armor plate at 100 meters, the Russian 14.5 mm antitank gun exceeded the usual values for such weapons. The German Wehrmacht used this weapon as the 14.5 mm Panzerbüchse 783 (r)-russ. PTRD 41.

Right: The Russian Light Machine Gun DP 28 had a good reputation among German grenadiers. It was designated 7.62 mm Light Machine Gun 120 (r)-russ. DP 28.

Below: Reports on the growing strength of British armored forces in North Africa (mid-October 1941) led to the hurried introduction of more capable antitank weapons, including the 7.62 cm Field Cannon 296 (r)-russ. 36- shown here.

This 3.7 cm Anti-Aircraft Gun 39 (r), shown here with its crew of Flak helpers, was among the numerous captured Russian guns. The picture was taken in the Halle-Merseburg industrial area in 1944.

Gunnery training with a 4.5 cm Antitank Cannon 184/1 (r)-russ. 37 by one of the eighteen Panzerjäger units of the Police. The Russian GAS-MM and ZIS-5 trucks were used to tow the guns.

This artillery unit was equipped with Russian cannon howitzers. The guns were designated 15.2 cm Cannon Howitzer 433/1 (r)-russ. 37-. The towing vehicles, American Studebaker US 6-U2 2.5-ton 6z6 trucks, were also captured from the Red Army.

A Light Artillery Tractor 601 (r)-russ. CT-3- tows the staff bus of an army unit through rough terrain. Large numbers of Stalinez tractors were also used for such purposes.

Only a few of these self-propelled gun mounts were used by a Panzerjäger unit of an infantry division. On the chassis of a 738 (r)-russ. T 26 tank, a 7.5 cm Antitank Cannon 97/38 was mounted. The barrel of this gun came from French supplies.

The many captured Russian Field Cannon 36 (designated 7.62 cm Field Cannon 296 (r)-russ. 36-) were reworked to take the ammunition of the German 7.5 cm Panzerjäger Cannon 40 L/46. The new designation was 7.62 cm Panzerjäger Cannon 36. In all, 363 of them were mounted on selected Czech Panzer 38 (t) chassis, beginning in April 1942. The resulting tank destroyer strengthened the antitank forces considerably.

The Weapons and Equipment Captured in Italy

In September 1943, the Kingdom of Italy withdrew from World War II as an ally of Germany. This development has been foreseen. German units disarmed the Italian Army in Italy and the Balkans. Thus further extensive supplies of weapons, ammunition, vehicles and equipment came into German hands.

The numerous Italian rifles taken over were to play a role a year later in the formation and arming of the German Volkssturm. The ammunition supplies for these weapons were so scant that often only ten bullets per rifle could be issued. Among these foreign weapons, the short and scarcely useful Italian carbines prevailed. The original intention of "boring out" the 800,000 available Italian carbines in Suhl "for normal bullets and equipping them accordingly" was rejected.

The Army often used Italian bolt mines. With the use of the O.T.O. Mod. 35 hand grenade and other models, the number of training accidents grew greatly, so that a special manual had to be issued. Italian hand grenades turned up in large numbers again at the end of April 1945, during the combat in Bozen. The local Volkssturm was armed with them. Italian grenade launchers were issued to the fortress of Küstrin (ten 8.1 cm Grenade Launcher 276 (i), with 300 rounds each) and the Dresden defense zone (49 4.5 cm Grenade Launcher 176 (i) with 300 rounds each).

The Italian artillery materials were outmoded; some of the older models came from the Czech manufacturer Skoda. Only a few guns, like the 15 cm Heavy Field Howitzer 404 (i)-ital. 149/19 M 37 performed up to the standards of the time. Among them, the 15 cm Cannon 408 (i)-ital. 149/40-, of which the Wehrmacht obtained fifteen, must also be noted.

A similar impression to that of the guns was made by the Italian tanks. Even so, the available materials were taken over and used, some by the Army divisions, some for Police tasks. For example, the 278th Infantry Division can be cited, as its Panzerjäger unit had a company of Italian assault guns. As can be read in the division's history, "the little Fiat vehicles" proved to be useless in combat. A memo from the 98th Peoples' Grenadier Division, of April 1945, indicates that this unit had an armored scout platoon with Italian armored scout cars. The only prototype of a Breda 2 cm quadruple anti-aircraft gun carrier ended up in the strets of the Markish town of Teupitz in April 1945. The vehicle has been taken by the V. SS Mountain Corps. According to a letter from the Chief of Army Armaments and commander of the Replacement Army, the Army Weapons Office had octained 84 Scotti and Breda 2 cm anti-aircraft guns in July 1944.

Italian motor vehicles and artillery tractors of all kinds were used in great numbers, the latter in the Panzerjäger units of several divisions, to tow antitank guns.

Italian assault guns were not popular; they saw service as of September 1943 with some divisions in Italy and the Balkans. 123 of the M 40 Assault Gun with 7.5 cm Cannon 75/18 850 (i) were on hand on October 1, 1943.

Captured American Weapons and Equipment

War materials from the USA were first captured by the Wehrmacht in quantities worth mentioning in North Africa in 1941 and on the eastern front in the winter of 1941-42. They included tanks, other vehicles and weapons. They came from the Lend-Lease supplies assigned by President Roosevelt to Great Britain on March 11, 1941 and to the Soviet Union that October. Between November 1941 and April 1942, the Soviet Union alone received, among other things, 2400 tanks, 189,000 field telephones, and 75,000 Thompson machine pistols. In 1942-43 the shipments of American motor vehicles amounted to as many as 10,000 a month. From these supplies came, for example, the 2.5-ton Studebaker US 6-U 2 trucks that some German units were already using in 1942. No weapons and vehicles came into German hands from direct contact with American troops until the 1943 combat in Tunisia and Italy. Their numbers remained small, coming nowhere near the numbers captured in the campaigns up to 1941, for the Wehrmacht was withdrawing then. Nothing changed after the opening of the second front in France in 1944. Captured war materials, as long as they did not have to be turned over to the Army Weapons Office for evaluation, were welcomed by the troops as reinforcements for their own weapons and vehicles. Requests by high offices to turn them in were obeyed unwillingly, if at all.

SS Standartenführer Otto Skorzeny was commissioned in the autumn of 1944 to establish a special unit, Panzer Brigade 150, for the Ardennes offensive that was being planned. The problems that arose from being equipped with American and British weapons and Vehicles are portrayed by Skorzeny in his memoirs. Two Sherman tanks, ten armored scout cars, two armored troop carriers, about thirty Jeeps and fifteen trucks were gathered. Twenty percent of the unit's specified number of American guns were lacking, and there was no ammunition for the antitank guns or grenade launchers.

During the winter combat of 1944-45 on the western front, the numbers of captured materials increased again; tanks, vehicles and handguns, and sometimes ammunition. This included 50,000 rounds of American 104 mm howitzer ammunition that fell into German hands in the Bastogne-Gouffalice area. Twenty thousand could be taken away. During the further combat, eighteen were fired from German field howitzers.

Another example: The 10th SS Panzer Division "Frundsberg" captured, among other things, twelve Sherman tanks near Herlisheim in Upper Alsace on January 17, 1945. They remained with the division until the war ended. In the night of May 7-8, 1945, they, along with the last German tanks, were blown up near Schmiedeberg in the Erzgebirge after a fight with Russian advance units.

This American M 3 A 1 troop carrier was captured by a German unit in the late autumn of 1944 and photographed in Alsace.

General Problems in Using Captured Materials

Capturing, transporting and gathering enemy war materials were governed by orders and decisions in the Wehrmacht. Rules were already made by the Wehrmacht High Command on June 16, 1940, and were later revised or clarified at various levels of command. Among them, surely as a result of experience in the sometimes chaotic conditions regarding captured materials from Poland, was a ruling on the possession of captured war materials.

In August 1940, a "reporting process to a new pattern" was prescribed. Even privately owned weapons had to be collected and handled as captured goods, and in an amendment from the campaign staff z.b.V.23, dated August 14, 1940, brass cartridge cases from World War I, found in many Belgian and French households, were added.

On May 27, 1943 the Wehrmacht High Command issued a new version of the rules about "booty and confiscation in already occupied France", which in principle was also valid for the other occupied lands.

In order to be able to define captured war materials, the troops used the already mentioned fifteen volumes of manuals for foreign equipment (D 50/1-15). Important for their reuse were also the likewise multipart operating manuals (D 51/2 Operating Manual for British Equipment of March 1, 1941, etc.). In addition, a flood of Army service manuals, messages and memoranda appeared. Nor was that enough. In the offices of the army groups and armies there was also printed information, such as the volume "Weapons of the Red Army", issued in 1944 by the Army Group Center/Ic/A.O.Evaluation.

The use of captured war materials in the German Wehrmacht can be categorized only with difficulty. Thus the attempt that follows must be regarded with reservations:

1. The immediate reuse of captured materials by the troops,

a. to cover existing shortages in vehicles, weapons or ammunition. This type of use of captured material was found everywhere, but remained limited to a short period of time.

b. The troops wanted to utilize already known advantages of enemy weapons for their own waging of war. One example is the use of Russian company grenade launchers and their ammunition in place of the German 5 cm Grenade Launcher 36.

2. The reuse of captured war materials on orders from the higher troop leadership, the Army High Command or the Wehrmacht High Command. This was generally preceded by a planned collection, examining and overhauling of the materials.

a. The rearming or German troop units with captured weapons, to be able to make up, at least for a time, for the high losses of German weapons and equipment.

b. The establishment of special troop units, which were equipped primarily with captured equipment. These included captured tank units, Army coast artillery units, police and security units. In isolated cases, they were also used to deceive the enemy.

c. The additional strengthening of divisions in line with weapons, for example, to cover personal losses.

3. The reuse of captured weapons and vehicles after a thorough reworking, carried on to make them suit German equipment or German ammunition. This was particularly true of armored vehicles and various artillery weapons, occasionally of handguns. Such modifications were often carried out by the industry, while others were made by the troops themselves.

4. The reuse by the Army Weapons Office, in order to determine combat capabilities, advantages and disadvantages. Naturally, this was also done to acquire information about the construction of the captured war materials. This was done in close cooperation with the industry. For example, the evaluation of Russian ammunition by Rheinmetall-Borsig at their Sömmerda works in February 1943.

5. The scrapping of captured war materials. This included the collection of the wrecks of shot-down aircraft, as well as the scrapping of grenade launchers and artillery. The Army High Command, in a memo of May 15, 1944, encouraged the "thoroughgoing scrapping" of more than 19,750 captured grenade launchers and guns. The large amounts of steel alloys thus gained were to be used in the manufacture of armaments.

The war materials captured in the campaigns between 1939 and 1941 in particular had a great significance for Germany. In comparison with its opponents, the German Reich had only limited reserves of raw materials and work forces. With these captured materials, the Wehrmacht, and especially the Army, could be expanded quickly. In the further course of the war, they helped to make up for developing shortages. The establishment of the German Volkssturm in the autumn of 1944 would not have been possible without captured weapons. But the use of numerous captured weapons and vehicles also caused many problems in the training and logistics of the troops, which very quickly set limits to further expansion.